Living in an Interconnected Universe

Modifying the vibrational flows of the universe for a better future

Michele Verny

Revive Publications

*A catalogue record for this book is available
from the British Library*

ISBN: 978-1-907962-37-0

Published by Revive Publications

Reading, England

For everyone & everything

Contents

Introduction

The universe is deeply interconnected; this is obvious. However, there are many different views concerning both the nature and the extent of this interconnectedness.

The view that the universe is deeply interconnected is sometimes associated with certain spiritual traditions, but it is now also firmly established by scientific evidence. Experiments in quantum mechanics conclusively show that the universe is deeply 'entangled' in ways that do not tally easily with our everyday interactions with the world. Or, perhaps I should say this evidence does not tally easily with our expectations concerning the nature of the universe which arise from our everyday interactions. According to the evidence two parts of the universe that exist at an immense distance from each other subtly affect and influence each other.

The purpose of this book is not to convince you that the universe is deeply interconnected – I presume that you already believe this. Rather our purpose is to explore the likely nature of this interconnectedness, and also to explore what it means for the decisions that you make in your life. Is there an appropriate way to act in a universe that is deeply interconnected? And, does the interconnectedness of the universe mean that we have a lot of power to change the universe? I suggest that the appropriate

answer to these questions is 'yes' and 'yes', and hope
that you agree by the time that you reach the end of
the book.

Living in an Interconnected Universe

What does it mean to live in an interconnected universe? And does living in a universe which is interconnected have implications for how one would like to live one's life? In other words, does the phenomenon of universal interconnectedness affect how one decides to act?

I presume that the person who realises that the universe is deeply interconnected will have an additional set of considerations at work in their decision making, and that they will therefore be likely to act differently. By this I mean that they are likely to act differently compared to how they themselves would have acted if they didn't have such a realisation; I also mean that they are likely to act in a way which is subtly different from a person who lacks this realisation.

One might expect that a person who realises the deeply interconnected nature of the universe would automatically realise the power that they have to alter the universe in a deep and pervasive way. For, if the universe is deeply interconnected, then one, as part of the universe, is deeply interconnected to large parts of the universe (if not to the entire universe). To realise that one is deeply intercon- nected is to realise one's power. In an interconnected universe one's actions will have a plethora of effects – many of these one will never realise, be able to fully appreciate or comprehend.

There are many different aspects to the phenomenon of interconnectedness. Let me start by considering two different types of interconnectedness – temporal interconnectedness and spatial interconnectedness.

Spatial interconnectedness is a phenomenon that exists between two parts of the universe in the present moment. It is reasonable to assume that there is a different level of interconnectedness between different parts of the universe in the present moment. By this I mean that it is reasonable to assume that I, as a living human being, am more interconnected to my clothes than I am to the planet Saturn. I spend a lot of time in very close proximity to my clothes – the more time we spend together the more we become interconnected. Contrarily, the planet Saturn is a great distance away from me, and always has been. It is surely plausible that, if the universe is deeply interconnected, that I am more interconnected to my clothes than I am to the planet Saturn.

This is plausible because it is reasonable to assume that there is a close link between spatial interconnectedness and memory. The more time that two parts of the universe have spent close to each other, the more familiar with each other they will be in the present. In other words, the parts of the universe have a memory within them which stores their past 'familiarity levels' with other parts of the universe.

What does this mean in practice? It means that almost all parts of the Earth will be deeply interconnected (a small minority of the parts of the Earth are recent additions to the planet, which have arrived as meteors from outer space). Furthermore, certain parts of the Earth will be more deeply connected than others. So, I will be more deeply connected to my clothes than I will be to the clothes of a person who lives in a distant country. I will also be very deeply connected to the people and the possessions that I spend a lot of time with.

I suggest that there is likely to be a simple correlation. The higher the 'familiarity level' that exists between two parts of the universe, the higher will be the level of interconnectedness between these two parts of the universe. What exactly is a 'familiarity level'? There will be two elements to this level – firstly, the closeness of two parts of the universe;

secondly, the length of time for which they have been close. So, two parts of the universe which have been very close together for a very long time will be much more deeply interconnected than two parts of the universe which have been very far apart for a very long time. However, there is a complex intricate web of relationships. So, it is possible for two parts of the universe which have never been closer than 10 metres from each other to have a higher 'familiarity level' than two parts of the universe which have been as close as 1 metre to each other. If the former parts were at a distance of 10 metres from each other for 10 years then they are likely to have a higher 'familiarity level' than if the latter parts had been 1 metre away from each other for 1 minute.

I hope that you can try and envision this great interconnected web of relationships existing throughout the universe at varying 'familiarity levels'. Perhaps it is useful to start with yourself. Think about the parts of the universe that you are most deeply connected to – the parts of the universe that you spend the most time with. Now, in your mind paint a picture of where you are in the present moment. When you have done this think about where all of the things are located that you are most deeply connected to (possibly your car which is in the car park, your clothes which are on your body and in your wardrobe, your bed which is in your bedroom, your golf clubs which are at the golf club, and all of the rest of your most used possessions).

Now, think about where all of the things are located that you are less deeply connected to. These are likely to include things such as the roads which you usually (or less frequently) travel along, the shops which you sometimes (or rarely) go to, and the places which you have been to on holiday.

You should attempt to have in your mind where you are in the present moment, and all of the different interactions / familiarity relationships you have built up with other parts of the universe throughout your life. Now, try and imagine the connections – the interconnections – that exist between you and all of these parts. Some of the lesser connections might involve a place that you went to only once 20 years ago, but try and imagine the link that exists between you and this place *in the present.* In your mind imagine the links – the 'rays' of familiarity radiating out from you to all of the parts of the universe which you are connected to at varying levels of deepness. Imagine that the 'familiarity levels' which are the most deep for you will involve radiating rays which are much more pronounced – much bigger and brighter – than those relationships which are less deep for you.

If you have achieved this then you will have in your mind some kind of idea of the way that you are interconnected to the universe. Furthermore, you will have in your mind which parts of the universe you are most interconnected to and which parts of the universe you are less connected to.

Is this just a picture in your mind or do these relationships actually exist in the universe? I believe that they actually exist. The phenomenon of home-opathy is grounded in the belief that such memory effects occur. There is also evidence that organs – such as hearts – which are transplanted from donor to recipient carry with them the memories of the donor. We all know that memories exist in the universe because we have memories within us of things that have existed in the past. Some people think that these memories are only stored in the 'brain' (or 'mind' if they believe this to be different from 'brain').

However, there are good reasons to believe that memories are not just stored in the brain but that they are stored in all parts of the universe – I have already mentioned that there is some evidence that memories can be stored in transplanted organs and in non-living parts of the universe (as revealed by the success of, and evidence for, homeopathy). So, if memories are stored in all parts of the universe, then there is every reason to believe that the differential memory relationships that I have outlined exist. It is only natural to believe that one will have a greater memory of that which one has a higher level of familiarity with. So, the rays of interconnectedness will be radiating out from you to all of your surroundings in a way that reflects your level of interconnectedness with these surroundings.

What I am calling spatial interconnectedness is linked to the phenomena in the world described by Chaos Theory. You might well have heard someone refer to Chaos Theory as the idea that a butterfly flapping its wings in one part of the planet can cause a tornado on the opposite side of the planet. In a deeply interconnected world very small things can have massive impacts, impacts that are unexpected due to both their nature and their scale. This is what I alluded to earlier when I referred to the immense power that humans have in an interconnected world – whether they realise it or not. If a butterfly flapping its wings can cause a distant tornado, just imagine the distant effects which can be produced by a group of humans acting in a particular way, or even a single human acting in a particular way:

Our actions create the world

There are a plethora of ways in which your actions create the world. Some of these ways will be obvious to you. You can see some of the effects that your moods, your actions, your body language, your good deeds, and your bad deeds, have on the parts of the universe that you are deeply interconnected with. However, just as a butterfly isn't aware that its actions have created a tornado, you are not aware of the multitude of effects that your actions lead to.

What kinds of effects are your actions likely to have? I suggest that there will be a strong correlation between your state of mind and your associated actions on the one hand, and the nature of the effects that radiate out from you to interconnected parts of the universe on the other. So, if you are in a kind, calm and peaceful state this would mean that you are radiating these states out to the parts of the universe that you are interconnected with. I should perhaps say now, as I haven't made it explicit yet, that many people believe that you might be deeply interconnected to the entire universe. I prefer not to think of interconnectedness in this universal way. Having said that, there are good reasons to believe, based on the theory of the Big Bang, that the entire universe is interconnected. My belief is that if the entire universe is interconnected this interconnectedness is likely to be a very very weak type of interconnectedness.

I prefer to focus on the strong bonds of interconnectedness that a person builds up with their surroundings throughout their life. This is a tangible sense of interconnectedness. One can make sense of this type of interconnectedness because one can see some of the effects with one's own eyes. I find it hard to make sense of the idea that my actions can affect distant galaxies at the other side of the universe. Of course, this doesn't mean that such affects do not exist. But if these weak and distant affects exist I am unlikely to change my actions – to live my life in a different way – due to the realisation that I might, at a very weak level, be interconnected to a distant galaxy at the other side of the universe.

I am interested in the deep and differential levels of interconnectedness that exist between you and your immediate surroundings. I believe that when one appreciates this web of interconnections it can make a profound difference to the way that one sees oneself, the way that one sees other people, and the way that one sees the other parts of the universe which one is deeply connected with. When you appreciate your power to change the world, and to develop deeper interconnections with particular parts of the universe, then this is a great event. What will do with your power? Which connections will you decide to deepen and which will you decide to weaken? The decisions that you make will not only affect your own future they will also affect all those parts of the universe which you are deeply connected with.

There is clearly the potential in such realisation to make positive changes which would benefit the human species; which would help to forge a future that you would like to see. The power is within us to make the interconnections change in the way that we desire.

I believe that most people, when they realise their power, will want to change the world into a more peaceful place with less crime, conflict and war. If enough people change then the ideal would be that these things (crime, conflict, war) are ultimately eliminated.

What is the cause of the crime, conflict and war that currently exists in the world? These phenomena currently are deeply embedded in the various interconnections that flow around the planet. These deleterious flows – the flows of crime, the flows of conflict, and the flows of war – are self-propagating. They have existed for so long that they have become deeply embedded and they are reinforced because they are radiating out from so many different humans. When a human is born on the planet today they are born into a web of interconnections which flow through them – and some of these interconnections are the deleterious ones (crime, war, conflict). Imagine a human born on a planet where there are no such deleterious flows surrounding them, and flowing through them, it is unlikely that they will start these deleterious flows themselves. They are likely to mould into the existing flows.

I assume that this is what will happen because I assume that the 'good flows' of interconnectedness are the underlying natural state of affairs. In contrast, the deleterious flows have been created by the group dynamics of humans. If this is right, then in the absence of a group of humans which produces deleterious flows, a new born human will not start deleterious flows itself.

The embedded deleterious flows of interconnectedness are hard to dislodge because they have existed for so long. However, the more humans that remove themselves from these flows the weaker they will become over time. One can either be a source which radiates out deleterious flows (flows of hate, of aggression, of conflict) and thereby propagate and sustain these flows on the planet, or one can simply be a source of 'good flows' – a source of peace, of positivity, of all that is good. The choice is yours.

I haven't made explicit what the phenomenon of 'temporal interconnectedness' is yet; although the interactions in the universe that the term refers to have been alluded to. Now seems like a good time to make explicit what the phenomenon is. Recall that 'spatial interconnectedness' is a phenomenon that exists between two parts of the universe in the present moment. In contrast 'temporal interconnectedness' refers to the past events that have shaped the present moment.

So, the state that one is in in the present moment (the state of one's body and the state of one's mind), is partially determined by the state that one was in in the moment before the present moment. There are other factors besides that state that one was in in the moment before the present moment that also determine the state that one is in in the present moment. One such factor is the phenomenon of 'spatial interconnnectedness' – the interconnections that exist in the present moment affect one's state in the current moment. It is also a possibility that in the present moment one can create new states within one – states that are solely new and wholly generated within one without being influenced by interconnections from either the past or the present.

Let us focus on how the states that one was in in the past influence the states that one is in in the present. The states that one was in in the moment before the present moment influence the states that one is in in the present moment. And the states that one was in in the moment before the moment before the present moment influence the states that one was in in the moment before the present moment. One can see how this regress of states will go back to the moment that was born. The states that were in one when one was born influence the states that one is in in the present moment. This is the phenomenon of 'temporal interconnectedness'.

There is much more to the phenomenon of 'temporal interconnectedness' than this – as you might well already have realised. The states that were in one when was born are clearly dependent on the states that regress back to the moment that one was conceived. And the states that existed in the moment that one was conceived are influenced by the states that existed in the moment before this moment. This regress continues backwards to the conception of one's parents. Similarly, the regression continues further backwards to the parents of one's parents, and to the parents of the parents of one's parents, and so on.

The regress clearly continues backwards to the first human that existed! So, to be clear, the state that one is currently in is partially influenced by the states that existed in the first human when this first human first came into existence. Of course, the states that existed in the first human when this first human first came into existence are influenced by that states that existed in the moment prior to the moment that the first human came into existence. So, the regress continues back through our non-human ancestors to the origin of life on Earth. The state that one is currently in is influenced by the state that existed in the first life-form to evolve on the planet when that life-form first-evolved.

As I am sure that you realise, the state that existed in the first life-from to evolve on the planet when that first life-from first evolved was influenced by the states that existed in the moment before the first life-form evolved on the Earth. So, the regress clearly continues back to the formation of the Earth, and further back to the formation of our universe – the Big Bang. The state that one is currently in is influenced by the state of the universe at the time of the Big Bang!

Clearly, if there was a universe that existed prior to the Big Bang which created our universe, then this 'prior' universe would influence the states that existed at the time of the Big Bang, and thereby influence the states that one is in in the present moment. According to the 'supercrunch' theory universes come and go – Big Bangs entail an expansion of the universe, which is followed by a 'collapse', and this collapse leads to another Big Bang.

So, the reach of the phenomenon of 'temporal interconnectedness' is immense. To say that the state that one is currently in is influenced by the Big Bang, and by past universes, if these pre-Big Bang universes exist, may sound slightly fruity. However, it is not as fruity as it might first appear. For, as with the phenomenon of 'spatial interconnectedness', there is a great variability in the strength of the differential interconnectedness relationships.

So, one's current state will be very greatly influenced by the state that one was in in the previous moment, and it will also be greatly influenced by both the state that one was in a few hours previously and the state that one was in a couple of days previously. In contrast, the state of the universe at the time of the Big Bang is going to have a negligible effect on the present state in one – it might even be better to conclude that this particular effect will be so weak that it is better to think of it as not existing.

There will be a vast array of temporal intercon- nections which link one with past moments of the universe – links between past states of the universe which have led to the states within you in the present moment. Some of these interconnections will be very powerful and strong, some will have moderate strength, and some will to all intents and purposes be non-existent; there is a whole gradation of interconnections of varying strength.

Try and picture these interconnections in your mind. Try and imagine how these links shoot through time – how they radiate out from you backwards into past states of the universe. These past states influenced the present state you are in at the moment. Rather than going backwards it is perhaps more helpful to imagine the states going forwards – imagine the moment that you were created in a zygote, the moment that you exited the womb and came into the world outside, imagine how the states that existed at these times led to new states (such as your childhood). Imagine the links as bright sparkly colours, as tangible rays moving through time, as new states of you are created by the movements of the rays from their past states. Have in your mind the whole progression of these sparkling flowing rays from the moment of your birth to the present moment. These rays and their progression through time influence your current state. You can change your future, you change the future, by changing the direction of these rays in the present moment. The current movements of the rays are determined by you (although they are influenced by the past and by other current states of the universe – i.e. by 'temporal interconnectedness' and by 'spatial interconnectedness'.) You make the rays dance and change and take on a new journey within the constraints – the channels – imposed by the influences from the past, and the influences from the wider universe.

Earlier you tried to imagine the interconnections in the universe that exist due to the phenomenon of 'spatial interconnectedness'. You have also just tried to imagine the interconnections that exist in the universe due to the phenomenon of 'temporal interconnectedness'. Now, try to imagine these two sets of interconnections simultaneously. What a wonderful picture must exist in your imagination! The rays radiate out from you to all parts of the universe that you are interconnected to at varying levels of deepness – both in the current moment and from the past states of the universe. It is hard to imagine states radiating out from you to other states of the universe (and to you) that currently exist ('spatial interconnectedness') whilst simultaneously imagining the states radiating to you from past states of the universe ('temporal interconnectedness') but it is worth the effort in trying to achieve such a vision. You will see the deeply related nature of the universe and the immense power that you have to shape the future to mould the future paths of the rays – to forge new interconnections – in both expected and unexpected ways.

Imagine how these interconnections become reinforced over time, how they become embedded in certain locations. You can imagine that these interconnections – the rays that radiate out from you and radiate through you – are certain patterns of vibrations:

> *The pattern of vibrations that comes from the universe to your body is modified by the pattern of vibrations that exits in your body as it moves through your body. You change the world by modifying these patterns*

Patterns of vibrations can become embedded as they become reinforced by similar patterns of vibrations. Imagine a house that due to some unexplained or strange occurrences becomes referred to as a 'haunted house'. The initial occurrence will lead to a particular 'spooky vibration' pattern in this particular house. As the reputation of the house as haunted spreads the people who visit the house to experience its haunted nature will be likely to have a set of expectations and feelings regarding the house before they arrive. They are nervous, scared, slightly excited, worried, and anxious as they approach, and step into and explore, the house. This particular vibrational state (the 'spooky vibrational' state) will radiate out from the people experiencing it to the house and be stored in the memory of the stuff of which the house is constructed. As more people explore the house two things occur. Firstly, the states within them get amplified by the states within the house. In other words, there is a low level 'spooky vibrational' state within the person as they approach the house, but when they enter and explore the house this state resonates with the pre-existing 'spooky vibrational' state that is already existing in the house and thereby causes an amplification of the 'spooky vibrational' state within the person to a high level state. Secondly, these states within the person reinforce the states within the house – the 'spooky vibrational states' become embedded within the house. There is a self-reinforcing process resulting from the initial labelling of the house as a 'haunted house'. The rays of vibrations that exist in, and flow in and through the house, become increasingly 'spooky'. When non-spooky rays of vibrations interact with the house (houses, just like people, have varying levels of both 'spatial interconnectedness' and 'temporal interconnectedness') then these rays become modified in a slightly 'spooky' way. What this means is that the haunted house is a location where 'spooky' vibrational patterns are centred – from this central location the patterns then modify and influence the surrounding universe in certain ways due to the interconnected nature of the universe.

What would happen if the house were to be demolished and its constituent parts widely distributed to different locations? The spooky vibrations would clearly still exist in all of the constituent parts and as the rays from the universe pass through these parts they would still be modified in a 'spooky' way. However, there would no longer be a central location which generates new 'spooky vibrations' which embed and reinforce these vibrations. This lack of the reinforcement means that over time the effect will fade, and the constituent parts of the house will becoming progressively less 'spooky' – their vibrational state will slowly change to suit the new pattern of interconnections which they are now exposed to, the new pattern of interconnections which they are now developing.

In the same fashion as 'spooky' vibrations become embedded in a haunted house various other types of vibrations become embedded in a multitude of other locations. 'Studious vibrations' become embedded in libraries, certain types of vibrations become embedded in hospitals, stock exchanges, graveyards, canteens, nightclubs, pubs, shops, leisure centres, everywhere.

Wherever a particular activity is carried out the patterns of vibrations of that activity will become reinforced in that location. These locations are hubs which home particular patterns, and due to the interconnected nature of the universe these hubs have significant affects on the wider universe – all of the parts of the universe which are significantly interconnected to the hub will have their vibrational state modified by the influence of the hub.

We can now envision the universe as a plethora of intermingling vibration patterns that radiate out from every point of the universe to interconnected parts of the universe – these rays flow through space and time creating a new universe as they flow. Hubs are locations of particularly intense vibrational patterns of a specific type which impart their vibrational state onto the flows which pass through them, and to the parts of the universe which they are interconnected to, in an above average manner.

All parts of the universe have particular vibrational states and play a part in creating the future states of the universe. What this means is that we can each play a part in creating the future that we would like to see. By being able to influence our own vibrational patterns we are able to modify the vibrational patterns that arrive to us from other parts of the universe, and we are also able to send out new types of vibration patterns to the universe.

Clearly, if enough people start sending out particular types of vibration patterns, and if enough people modify the vibrational patterns that reach them from other parts of the universe, then, over time, the impact will be significant.

In an interconnected universe we are all powerful agents of change. We all have the ability to modify the vibrational patterns of the universe – to change them into more desirable patterns

The vibration patterns of anger, jealousy and hatred – the particular patterns of vibrations which exist in, and flow through, various parts of the universe – can gradually be modified into more peaceful and harmonious rays.

By imagining the interconnected flows of vibrations which pervade the universe, and ideally by 'tuning into' these flows – by literally feeling them, feeling the patterns which have become embedded in particular locations, one can change the universe for the better.

If one is able to 'tune into' the flows of vibrations then one will also be able to surround oneself with the flows that one finds to be good for one, and avoid the flows that make one feel uncomfortable. This can be expected to be good for your health, as well as engendering an insight in to the vibrational nature of the flows of the universe – flows which are permeated by temporal and spatial interconnectedness.

I don't know if you are clear in your mind what these flows are and how they are related to the interconnectedness of the universe. Imagine a particular ray of sunlight as it leaves the Sun and travels to the Earth. This act of travelling is a flow that exists in the universe – a flow is simply a movement from one place to another. The universe is a summation of flows. And flows are summations of vibration patterns. Therefore, the universe is a summation of vibration patterns. As the ray of sunlight moves from the Sun to the Earth – flowing with its particular vibration patterns – its vibration pattern is modified due to the interconnected nature of the universe. The flow itself has a plethora of interconnections with the rest of the universe – both temporal interconnections and spatial interconnections – and these interconnections mean that the vibrational pattern of the sunlight is changed as it travels. This is simply a single instance of that which pervades the universe.

A constant flow of flows modified by spatial and temporal interconnections

Remember that you are part of the flows – you modify the flows, you create new flows. If you change yourself, then you do not just change yourself:

> *If you change yourself not only do you send out new patterns of vibrations, you also modify all of the vibrations that pass through you*

Seek to change your vibrational patterns – the vibrational patterns that exist within you – to become harmonious – become a harmonious hub of vibrations which are in alignment with the positive forces (the positive rays/vibration patterns) flowing through the universe. Embrace your power to improve not only your life, but to improve the universe too – the results will be amazing.

You will almost certainly not be aware that you are responsible for many of the joyous things that occur in the universe due to you changing your vibration pattern. However, because of the interconnected nature of the universe, you can be sure that a multitude of positive events, a multitude of joyous changes have occurred, and will continue to occur in the future.

Other books from Revive Publications:

Understanding Fear: The key to a brighter future

Dick Stoute   (2011)

Japan Quake: Why do humans live in dangerous places?

Simon Saint   (2011)

www.ingramcontent.com/pod-product-compliance
Lightning Source LLC
Chambersburg PA
CBHW071326200326
41520CB00013B/2875